Beautiful Stars
For Kids

Nature Books for Kids
By K. Bennett
Mendon Cottage Books

JD-Biz Publishing

Download Free Books!
http://MendonCottageBooks.com

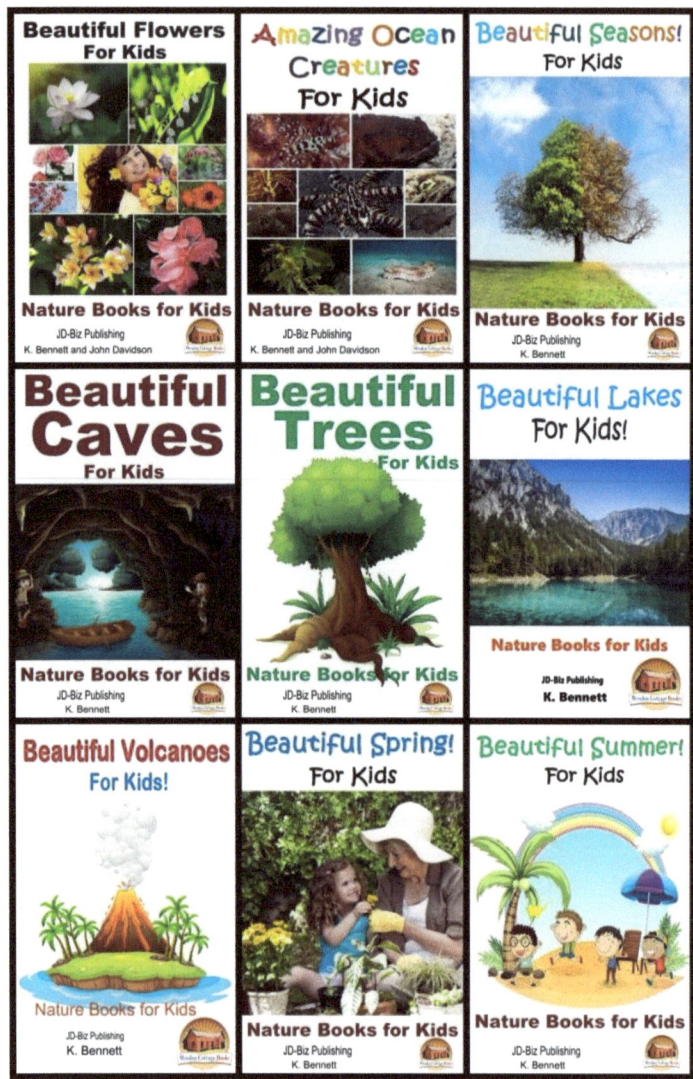

Purchase at Amazon.com

Download Free Books!

http://MendonCottageBooks.com

Table of Contents

Introduction 4

Chapter 1: Types of Stars 9

Chapter 2: Star Galaxies 15

Chapter 3: Fun Facts! 22

Become a Space Explorer! 25

Vocabulary: 31

Conclusion: 32

Sources: 34

Author Bio 35

Publisher 39

Introduction

*Be humble for you are made of earth. Be noble for you are made of stars. ~ **Serbian Proverb***

Look up at the night sky. Do you see bright lights? They might look blue, red, orange, or white. What are they? Stars? Yes! Good job.

The universe is full of twinkling balls of gas and energy that shine!

But did you know not everything you see is stars? That's right! Some of those lights might be planets. Others might be satellites racing across the night sky.

Many people love to look up and wonder… what's happening up there? How about you? Do you wonder what's floating around in space?

There are lots of things we can find like: galaxies, planets, comets, asteroids, moons, meteorites, and more.

But did you know stars are a big part of the space? Let's learn more about these amazing balls of light.

What is a star?

If someone asked you that question, what would you say? A great answer is found at ***Easyscienceforkids.com***. It says:

"Stars are giant balls of gas. Clouds of dust and gas swirl through the universe. Sometimes this dust and gas begins to collect in one area. As more dust and gas collect, the mass becomes heavy. It starts to swirl and becomes hot. When it gets really hot, it begins a process known as nuclear fusion. If this mass gets really big and hot, it becomes a star."

Stars are amazing balls of energy and they live for a long time. How long? Some stars are billions of years old! And they live busy lives. Stars turn hydrogen into helium all the time. This energy is used every day to keep stars alive. When the energy runs out, stars cool down. First, it might turn white or blue. Then it will turn orange or red. When the energy is gone, the star will turn dark or black.

Stars can blow up after many years. When they do, it's called a Supernova. In our book *Beautiful Black Holes for Kids*, we explained what happens when stars explode.

Do you remember?

What is a supernova?

The core of the star must change for a supernova to happen. NASA says we have two types of supernovas:

The first is a binary star system. Binary means two stars. They orbit each other like good friends! One is called a white dwarf. This star is a sneaky little ball. Did you know it steals matter from the other star?

Yes, it does!

But when it gets too much energy, the white dwarf has nowhere to put it and can't hold it anymore. Can you guess what happens next? That's right! It explodes into a supernova.

The second is when a star grows old. After it gets too old it's ready to die. The fuel the star needs to keep alive runs out, like a car with no gas.

When the fuel is gone, the mass of the star leaks into the core. When it gets too heavy to hold the mass, the core collapses. But the star doesn't die quietly. It goes out with a big, loud bang! Then a giant explosion rips through space and a supernova is born!

What else can we learn?

The North Star is very special star called Polaris. Unlike other stars that move around in the sky, it seems to stay in one place! This is a good thing. Do you know why?

Polaris is used by sailors and navigators to find their way home. This is star is also called the guiding star. Lots of other people use this star to travel.

Learning about the planets can be lots of fun, but learning about outer space can be lots of fun too. Welcome to the amazing universe of…
Beautiful Stars!

Chapter 1: Types of Stars

Stars are amazing objects in the night sky. Just like humans and animals, some are big and some are small. They have their own personalities too. Let's learn how some stars differ from other stars.

We will use the sun as an example. What do you know about the sun? For one thing, the sun is a star and not a planet. Can you tell the difference between a star and a planet? This short list will help you.

Stars

Stars look like they twinkle in the night sky.
Stars make their own light.
The gas in stars is called plasma. This makes stars very, very hot.
Stars have lots of mass and gravity.

Planets

Planets don't look like they twinkle, but they shine.

Planets can't make their own light. They reflect the sun's light.

Planets are not as hot as stars.

Planets have less mass and less gravity.

Now you know the difference between stars and planets. Let's learn about the different kinds of stars.

Are you ready? Great!

There are seven types of stars. They are known by the letters O, B, A, F, G, K, M. Can you remember these letters? Brainstorm and come up with a way to help you remember!

Stars are special

Each star has a color.

-O are blue stars.

-B are blue-white stars.

-A are white stars.

-F are yellow-white stars.

-G are yellow stars.

-K are orange stars.

-M are red stars.

-O, B and A are hot and burn brighter than the sun, they are approximately 7,500 to over 25,000 K.

-F stars are 6,000 – 7,500 K.

-G stars are 5,000-6,000 K.

-K stars are 3,500-5,000 K.

-M stars are under 3,500 K.

The **K** stands for Kelvin. This is a way to measure the temperature of heat. If you want to know the temperature in Fahrenheit, just apply it to this formula:

$$°F= (K-273.15) \times 1.8000 + 32.000$$

Stars big and small!

Did you know some stars have funny names like dwarfs and giants? Yes, they do!

White dwarf

A big part of this star is made of carbon. These stars are about the size of earth but much, much heavier!

Yellow dwarf

These stars burn their energy fast. Our sun is a yellow dwarf. These stars live for a long time. Astronomers say yellow stars live for 10 billion years! That's a long time!

Red dwarf

These stars live much longer because they burn or use their energy slow. They can live for trillions of years before they run out of energy! This kind of star does not shine very bright.

Blue giant star

This star is a big and blue. It is very hot and burns helium. This star burns so hot it lives for a much shorter time than other stars. When these stars die they go out with a big kaboom! Scientists use blue stars to find newborn stars in space.

Red giant star

This star is a big, old star. It burns hydrogen and when it runs out of energy is turns into a red giant with an orange color.

Super giant star

These stars are huge. They are so big they can fill our whole solar system! These stars are hard to find and when the die, they explode into giant supernovas and turn into big, black holes!

Black holes are dangerous. In our book *Beautiful Black holes for kids* we explained the reason:

"The force of a black hole is so strong light cannot escape. Do you know what happens to light when it gets near a black hole? Strong

---

gravity pulls light and everything else into the center. It is so strong that nothing escapes the powerful force, and everything falls inside!"

QUICK FACT FOR KIDS:

Have you heard the song: "Twinkle, twinkle little star?" Did you know stars don't really twinkle?

The reason it looks like its twinkling is because light comes through atmosphere. When the light moves through the air and the wind in the layers of the earth, the light of the star is bent into different parts of light. Then the light spreads all over the place. This makes the star looks like its twinkling.

Chapter 2: Star Galaxies

There are other types of star formations in the universe. Did you know there are star nurseries where stars are born all the time?

Stellar nursery:

The nebula (one) or nebulae (more than one), is a stellar nursery where lots of big, hot stars are born.

Stellar nurseries are full of hydrogen and helium gas in a huge place. Some nurseries are light years across from one end of the nursery to the other end. That means it's trillions and trillions of miles wide!

Nurseries have interesting shapes. Some look like clouds, and others looks like horses. Some even look like crabs!

No one knows exactly how star nurseries are formed, but they are very important in making more baby stars.

Galaxies:

Galaxies are made of beautiful, bright stars. There are many galaxies in the universe. It is hard to know how many galaxies are in space but scientists say there are billions. And each galaxy holds billions and billions of stars. Isn't that amazing?

Galaxies have names of their own like spiral, elliptical, and irregular. Each one of these galaxies is special.

1 - Spiral

These galaxies spin in a spiral shape. There are 3 parts to a spiral galaxy. They have a bulge like a ball in the center of the galaxy. The other part is a disk, and the last part is halo. The halo is around the ball in the center.

Older stars are in the center of the galaxy. Younger stars, dust, and gas are found in the disk part. And in the halo part we find stars in a globular cluster. This just means groups of stars together.

The interesting thing about the disk part of the galaxy is the arms. Yes, the galaxy has arms just like you do! Of course, it looks different than human arms, but it spins around the galaxy giving it a neat shape! This is also where stars are formed.

The universe is a busy place!

2 - Elliptical

These galaxies look like a big oval. They can be small or big. New stars are not formed in elliptical galaxies. This means the stars are old. The stars are so close to each other in this kind of galaxy that the center looks like one big, giant star!

This galaxy is so bright that if planet earth was inside it, we would have bright lights both day and night. What do you think about that?

3 - Irregular

This just means a galaxy that is not spiral or elliptical and has an interesting shape.

Now that you've learned about the types of galaxies, let's learn about three of the most important galaxies we know!

The Milky Way Galaxy

This galaxy is where we live; it's a spiral galaxy and looks a little milky. This is why it's called the Milky Way.

Our galaxy is full of stars. Scientists say there are more than 100 billion. Some say it's 400 million.

Wow…that's a lot of stars!

Andromeda Galaxy

This galaxy is not too far away. Andromeda is a spiral galaxy and we can see it at night if we look up into a clear sky.

Andromeda is full of stars... more than 1 trillion! That's a lot more stars than the Milky Way.

The Local Group

This is a cluster of galaxies with approximately 30 galaxies. The Milky Way galaxy, Andromeda, and Triangulum are a part of this group.

DID YOU KNOW?

Proxima Centauri is the nearest star to our planet; it is 4.2 light years away. Remember: light travels fast! It is the fastest speed we know in the universe.

If there's nothing to slow it down or stop its speed, light travels at 186,282 miles per second! If you prefer kilometers, it's 300,000 thousand kilometers per second. How fast can you move?

The biggest star that we know about is UY Scuti. How big is it? Did you know it would take 1,708 suns to fit inside the red giant? That's a big star!

What about the earth? It would take 6.6 quadrillion earths to fit inside UY Scuti! Isn't that amazing?

Why don't you do more research? I am sure you will find lots more information on stars!

Chapter 3: Fun Facts!

I hope you are enjoying this book on Beautiful Stars! Here are a few more neat facts:

- Stars are a lot like humans. They are born, live for a while and die, but it takes a long, long time. Some stars live for thousands and billions of years. But when they die, a few of them blow up and go supernova. Others do not blow up. They become white dwarfs and soon fade into space. Others become black dwarfs.

- Many years ago, people looked up at the sky and saw different shapes or patterns. These people thought the shapes looked like animals, humans, or Gods. We call these patterns constellations. There are many constellations in space. The International Astronomical Union has a list of 88 constellations.

- It takes a long, long time for the light of stars to reach us on planet earth. Sometimes it takes billions of years. So when you look at the stars, you are looking back in time!

-At night, we can only see between 2,500–3,000 stars with our eyes.

-Globular clusters are full of bright stars. This word comes from a Latin word that means "a small sphere." These stars are some of the oldest stars in space. Omega Centauri is the brightest globular cluster in the Milky Way Galaxy.

- When a normal star dies, it gets rid of its outer layers. After a while a nebula is formed. This star formation is beautiful and one of the most famous nebulas we know is called: sunflower nebula or Cat's eye nebula. And yes… It does look like a cat's eye. What do you think?

- Star comes from a Greek word that means "aster."

-Scientist continue to add new information to their theories on the universe. There's lots they don't know but they're still searching!

- Big stars do not live very long. Can you guess why? They use up their energy too fast! Smaller stars live for much longer.

- Cassiopeia, Orion, Phoenix, the Great Bear, and Lynx are the names of a few popular constellations.

Become a Space Explorer!

This project is adapted from ***mykidsadventures.com***. This is simple and fun to do. Here are the steps:

Go outside on a clear night. Look up at the night sky. What do you see? Make a list in your mind or on a piece of paper if you prefer.

The next night, do the same thing. Do you see anything different? Try to look at the night sky for at least one week, each night. Can you see the changes?

Each night the sky is different. You might notice the moon is smaller some nights and larger on other nights. A star might appear to twinkle really bright one night and the other night it might be dimmer. You might even see a shooting star!

Another fun project

Making constellations can be fun…and if you make them with marshmallows you can enjoy a sweet treat and learn about the stars at the same time. Don't forget to get permission before you try this project.

This is what you will need:

-1 bag of small marshmallows (mini)

-Toothpicks

-Crayon or pencil (White)

-Black construction paper (This is your night sky)

-Diagrams of constellations. You can find these online or in books.

-A moon chart. You can also find this online or in books.

-If you have an Apple or Android device, you can find the Apps in the store. Do not install these apps without your parent or a guardian's permission.

-You can use binoculars or a telescope for more fun!

After your supplies are ready, get started. First, read the information on the constellations you want to learn about. Choose one of them. Then draw the constellation with the white crayon or pencil on the black construction paper.

Use the pictures of the constellation to guide you. Then build the constellation using marshmallows for the stars and toothpicks for the imaginary lines.

 Have fun!

One more idea

Stars, constellations, nebulas and star nurseries are great choices for science experiments. If you use this topic, don't forget the steps you need to make it a great project.

Steps:

1 – You need to ask a **question** to be answered by observation or experimentation. Make it a very interesting question so your classmates and teachers will want to learn the answers!

For example: What will a star look like if you could get close? What would it feel like if you could get inside the core? How hot would it be? Can you live there? Why or what not? Is it different from earth? Make a list and find ten reasons why a star is different from planet earth.

2 – The next step is to state a **Hypothesis**. This is a big word but Sciencekidsathome.com explains it like this*: It is a tentative*

explanation for an observation, phenomenon, or scientific problem that can be tested by further investigation.

Your hypothesis is what you think the results of your project will be when your research is all done!

Write your ideas on what you think you might discover on stars.

3 – Next on the list is: **Procedure.** This is very important. Procedure will help you discover the answer to your question and prove what you are trying to say.

There are other experiments online that can help you. Ask your parent or a guardian to help you search. Or ask for permission before you search.

4 – **Results**. You will need to show your results and all the information you collected for your project.

5 – **Conclusion**. Finish up with what you learned and then answer the question you had in Step 1. If you can't answer the question, explain why the question cannot be answered.

I know you will have fun learning about the universe and all its wonders!

If you don't like the ideas in this book, put on your thinking cap and come up with your own conclusions! I am sure you will do an amazing job!

Space Quiz

Quizzes can be a fun and exciting way to learn. This quiz is adapted from **Enchanted Learning**. Put on your thinking cap and have fun! Don't forget to get permission from a guardian or your parents before searching online!

1 – What was the first animal sent to space?

2 – Who was the first person sent to space and when did he go there?

3 – Who was the first American sent to space and when did he go there?

4 – Who was the first woman sent to space and when did she go there?

5 – Who was the first man to orbit planet earth and when was he sent into space?

6 – What is the name of the astronaut who spent the most time in space?

7 – What is the name of the first person to walk on the moon?

8 – What is the name of the second person to walk on the moon?

9 – On what day did people walk on the moon for the first time?

10 – What is the name of the first space station and when was it launched?

Find the answers… in the conclusion!

Vocabulary: Stars are amazing objects in space! Here is a small list of vocabulary words to help you learn more.

-Absolute zero

-Accretion disk

-Asteroid

-Antimatter

-Aurora

-Binary

-Blue moon

-Black moon

-Celestial sphere

-Chromosphere

-Doppler effect

-Dwarf planet

-Event horizon

-Fireball

-Galactic nucleus

-Galilean moons

-Globular cluster

-Heliosphere

-Interplanetary magnetic field

-Kilo parsec

-Luminosity

-Magellanic clouds

-Neutrino

Do you know what these words mean? If you are not sure, ask your parent or a guardian's permission to search for the definition. I hope you learn something new!

(*www.dictionary.com*)

Conclusion:

Our universe is full of wonderful things and stars are an amazing part of it. We only see a small number when we look up at night but there are millions of bright balls shining down at us. And even though we don't know everything about them, let's keep on learning. Imagine what amazing things we will discover one day!

We hope you have enjoyed this book on Beautiful Stars. Remember…

"Educating the mind without educating the heart is no education at all." - *Aristotle*

Happy Learning!

Space Quiz Answers

1. A dog

2. Yuri Gagarin; April 12, 1961

3. Alan Shepard; May 5, 1961

4. Valentina Tereshkova; June 16, 1963

5. John Glenn; February 20, 1962

6. Sergei Avdeyev

7. Neil Armstrong

8. Buzz Aldrin

9. July 20, 1969

10. Salyut 1, April 19, 1971

Sources:

http://easyscienceforkids.com/all-about-the-stars/

http://www.kidsastronomy.com/stars.htm

http://www.enchantedlearning.com/subjects/astronomy/stars/twinkle.shtml

http://www.enchantedlearning.com/explorers/spacefirsts.shtml

http://www.campliveoakfl.com/site/14-fun-facts-stars-get-kids-excited-astronomy/

http://www.mykidsadventures.com/discover-astronomy-for-kids/

Author Bio

K. Bennett loves to write for both children and adults. Many subjects are interesting to research, but writing for children is special to her heart.

Her favorite pastimes include reading, traveling and discovering new things. Each of these activities helps to fuel her imagination and acts like a blank canvas waiting for more stories.

She is intrigued with fantasy elements like hidden worlds and faraway lands. And basically anything that gets her imagination soaring to new heights!

Her writing credits include children books online, short stories for online magazines, and novellas listed at Amazon.com

Our books are available at

1. Amazon.com

2. Barnes and Noble

3. Itunes

4. Kobo

5. Smashwords

6. Google Play Books

Download Free Books!
http://MendonCottageBooks.com

Publisher

JD-Biz Corp

P O Box 374

Mendon, Utah 84325

http://www.jd-biz.com/